Can We Feed Ourselves?

A Focus on Asia

To Cornell and Edie Capa
whose human concerns have inspired me.

Can We Feed Ourselves?
A Focus on Asia

Photographs by

Hiroji Kubota

Foreword by Lester R. Brown

Introduction by Gordon R. Conway

Magnum Photos

Published by Magnum Photos, Inc.
151 West 25th Street
New York, NY 10001-72004
Tel: 212.929.6000
Fax: 212.929.9325
Email: photography@magnumphotos.com
Website: magnumphotos.com

Paris
Tel.: 331.5342.5000
Fax: 331.5342.5002

London
Tel.: 44207.490.1771
Fax: 44207.608.0020

Tokyo
Tel.: 8133.219.0771
Fax: 8133.219.3088

Design and Layout: Tomiyasu Shiraiwa
Printing Direction: Tomoyuki Wada
Reseach Editor: Sheryl A. Mendez
Jacket Design: Tomiyasu Shiraiwa and Eleen Cheung

Printed in Hong Kong by DNP

ISBN: 0-914122-00-2

This book was originally published in Japanese by Ienohikari Kyokai (Publishers),Tokyo, in November 1999, with a foreword by Lester R. Brown, President of Worldwatch Institute.

FOREWORD

"Can Asia feed itself?" is one of the central questions that will shape the human prospect during the twenty-first century. Of the world's population of 6 billion, 3.6 billion—well over half—live in Asia. China, the world's most populous country, contains nearly 1.3 billion people. India has just reached 1 billion. During the next half century, China's population is expected to grow by 200 million and India's by more than 500 million.

The environmental devastation of Asia is affecting the food prospect. Deforestation is leading to more destructive flooding. With 85 percent of the original tree cover of China's Yangtze River basin gone, the 1998 flood of the Yangtze River basin was one of the world's worst of this century. Flooding in China in 1998 drove an estimated160 million people from their homes.

The list of problems appears endless: population explosion, water shortages, environmental destruction, illiteracy, economic crisis, food shortages, and, in the case of North Korea, famine. Hiroji Kubota has focused the lens of his camera and his talent as a photographer on the food situation in Asia.

Kubota, a member of Magnum Photos, has been documenting Asia for many years. He felt so compelled to make the Asian situation known to the world that he painstakingly completed this splendid book of photographs, "Can We Feed Ourselves?"

In this project he focused on food, but one cannot look at Asia's food prospects without looking at the population pressures. With the cropland area in Asia no longer expanding, more people means less cropland per person. At some point, as the cropland area shrinks and as the demand for food expands, the growing demand for food can be solved only by turning to the outside world for imports.

Japan, South Korea, and Taiwan already import more than 70 percent of their grain. It is one thing for a country of 125 million people, like Japan, to import more than 70 percent of its grain from the outside world, but quite another thing when a China, with 1.3 billion people, turns to the outside world for a large share of its grain supply. Then the question becomes not only "Can Asia feed itself?" but "Can the world feed Asia?" This is the second key question.

Perhaps the greatest single threat to food security in Asia is spreading water scarcity. Water tables today are falling in China almost everywhere the land is flat. Under the North China Plain, the water table drops by 1.5 meters per year. The Yellow River, the cradle of Chinese civilization, now runs dry for a large part of each year.

The water situation may be even more dire in India. The latest analysis of India's water situation indicates that the amount of water pumped from underground is double the rate of aquifer recharge. As a result, water tables are falling throughout much of the country. The International Water Management Institute, based in Sri Lanka, estimates that aquifer depletion could reduce India's grain harvest by up to one fourth. In a country adding 18 million people per year, this is not a pleasant prospect.

Falling water tables in India could lead to soaring food prices. Unless the government can effectively manage spreading water shortages, the loss of confidence in the government could lead to political instability and severe food shortages. With 53 percent of all children undernourished and underweight, life for most of the 1.3 billion people living in the Indian subcontinent is precarious even now.

Not every face of every person in every photo of Hiroji Kubota's collection has an expression of poverty or suffering. The lives of people who struggle daily with insurmountable obstacles are captured in these photos. They reflect not only scenes of poverty and despair but also radiant moments of joy and hope.

Lester Brown
President, Worldwatch Institute

Food for All in Asia—Can It Be Done?

Nature can be bountiful. The ricefields of Asia, so evocatively photographed by Hiroji Kubota, have fed people and sustained civilizations for thousands of years.

But natural bounty by itself is not enough. It has to be matched with human ingenuity. The Green Revolution, a joint product of Western and Asian science and technology, was one of the great success stories of the second half of the twentieth century. Food production in the developing countries kept pace with population growth. Since 1960 an additional three billion people have been fed. Without the Green Revolution many of these, perhaps as many as two billion, would be starving.

Yet, like most technological advances, it was not a complete success. It had its drawbacks. There is still hunger in the world. Hiroji Kubota captures the reality of hunger today and also, more subtly, the prevalence of malnutrition and the constant struggle facing so many people in their daily quest for enough food for themselves and their children.

Today, about eight hundred million people, or some 15 percent of the world's population, get less than 2,000 calories per day and live a life of permanent or intermittent hunger and are chronically undernourished. Over five hundred million of these people live in Asia. Many of the hungry are women and children. More than 180 million children under five years of age are severely underweight (150 million in Asia). This represents a third of the under-fives in the developing countries. Young children crucially need food because they are growing fast and, once weaned, are liable to succumb to infections. Seventeen million children under five die each year and malnourishment contributes to at least a third of these deaths.

Lack of proteins, vitamins, minerals, and other micronutrients in the diet is also widespread. Over one hundred million children suffer from vitamin A deficiency. As has been long known, lack of this vitamin can cause eye damage. Half a million children become partially or totally blind each year, and many subsequently die. And, as recent research has shown, lack of vitamin A has an even more serious and pervasive effect, apparently reducing the ability of children's immune systems to cope with infection. About two million children die each year as an indirect effect of vitamin A deficiency.

Iron deficiency is also common in the developing countries, affecting a billion people. Over four hundred million women of childbearing age (fifteen to forty-nine years old) are afflicted by anemia caused by iron deficiency. As a result they tend to produce stillborn or underweight children and are more likely to die in childbirth. Anemia has been identified as a contributing factor in over 20 percent of all postpartum maternal deaths in Asia and Africa.

Paradoxically, hunger is common despite twenty years of rapidly declining world food prices. Although in many developing countries there is enough food to meet demand, large numbers of people still go hungry. Food prices are low, yet they remain high relative to the earning capacity of the poor. Market demand is satisfied, but there are many who are unable to purchase the food they need and, hence, to them the market is oblivious.

Not surprisingly, hunger is closely related to poverty. To the casual observer, poverty seems to be worse in the cities but, in reality, the urban poor fare better. About 130 million of the poorest 20 percent of developing country populations live in urban settlements, most of them in slums and squatter settlements. Yet 650 million of the poorest live in rural areas, often in the midst of healthy fields of grain and gardens of fruits and vegetables. But if they do not have the land or the income from employment, the food is out of their reach, and they are as malnourished as any urban slum dweller.

The first question we ought to ask ourselves is, why should we be concerned? Probably everyone who reads this book and looks at these photographs is getting an adequate diet. Does it matter to us that others are not so fortunate? Does it matter to the industrialized countries that many people in the developing countries are malnourished? Part of the answer to these questions is political. The end of the Cold War has not brought about an increase in global stability. While conflict between East and West has declined, there is a fast growing divide between the world of the peoples, countries, and regions who "belong" in global power terms and those who are excluded. Yet this potentially explosive inequity receives relatively little attention in the industrialized countries. The volume of agricultural aid going to developing countries is stagnating in real terms. We need to recognize that unless the developing countries are helped to realize sufficient food, employment, and shelter for their growing populations or to gain the means to purchase the food internationally, the political stability of the world will be further undermined. In today's world, poverty and hunger, however remote, affect us all.

At the same time, the growing interconnectedness of the world—the process commonly referred to as globalization—holds the promise of alleviating, if not eliminating, poverty and hunger. Globalization while threatening, on the one hand, to concentrate power and increase division, on the other contains the economic and technological potential to transform the lives of rich and poor alike. Much depends on where our priorities lie and, in particular, whether there is sufficient access by the poor to the economic opportunities created by the products of the new technologies.

Prospects for the Year 2020

If nothing new is done, the numbers of poor and hungry will grow. Most populations in the developing world are still increasing rapidly. By the year 2020 there will be about an extra 1.5 billion mouths to feed. If the proportion of the population of the developing countries deprived of an adequate diet remains the same, the number undernourished twenty years from now could be well over a billion.

What is the prognosis for feeding the world's population in the twenty-first century? Producing forecasts of world food production is complicated. Some predictions are relatively optimistic. They claim that the production of the major food crops will meet the demand over the next twenty years. Thus, in theory, world population growth rate will be matched by a similar growth in food production and food prices will continue to decline. Nevertheless, the developing countries as a whole will not be able to meet their market demand. In the IFPRI (International Food Policy Research Institute) model the total shortfall by 2020 is some 190 million tons, which will have to be imported from the developed countries. Asia will be importing some forty million tons.

Inevitably, models of this kind raise more questions than they answer. Most important, the food needs of the poor and hungry are omitted. As in the real world, they are simply priced out of the market and their needs are "hidden." By 2020 the total number of malnourished children will have declined slightly to 155 million, but over 100 million of these will be in Asia. And, probably, there will still be close to three-quarters of a billion people chronically undernourished.

The Decline in Yield Growth

These models also make optimistic predictions about crop yields and production. But, there is evidence, albeit largely anecdotal, of increasing production problems in those places where yield growth has been most marked. For example in the Punjab of India (one of the centers of the Green Revolution), wheat production is now being seriously threatened. Of greatest concern is the growing scarcity of water. According to several estimates, good quality water availability in the state is about 25 million acre feet, but the demand of the existing cropping intensity is about 37 million acre feet. There are some three-quarters of a million tube wells drawing water at greater than the recharge rate. In the most intensively cultivated districts the ground water table has fallen to a depth of nine to fifteen meters and is falling at about half a meter a year. Salinization is also serious, affecting 9 percent of the total cropped area, as is waterlogging. This and other, albeit largely anecdotal, evidence from the intensive grainlands of Asia, for example Luzon in the Philippines and Java in Indonesia, suggest there are serious and growing threats to the sustainability of the yields and to the production of the Green Revolution lands.

There is also widespread evidence of declines in the rates of yield growth. In the early years of the Green Revolution yields of rice, wheat, and maize were growing at thirty to forty kilograms per hectare per year. At the peak of the Green Revolution, in the late 1970s and early 1980s, yields were increasing at between sixty and eighty kilograms per hectare per year. But since the late 1980s the increases are down to around forty kilograms per hectare per year.

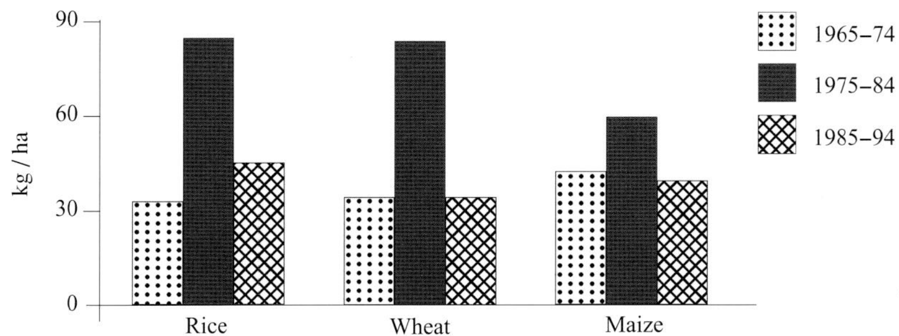

Average annual increase in developing country cereal yields by periods.

A combination of causes is responsible. In parts of Asia declining prices for cereals are causing farmers on the best lands to invest more in higher value cash crops. But, more important, there has been little or no increase in yield ceilings of rice and maize in recent years. A third factor is the cumulative effect of environmental degradation, partly caused by agriculture itself. Virtually all long-term cereal experiments in the developing countries are showing marked downward trends in yields.

Agriculture and the Environment

Environmental degradation is a sign of the failure of human ingenuity. Increasing agricultural productivity is not enough. The increase has to be sustainable, and that is the greatest challenge to our ingenuity.

The litany of environmental loss is familiar. Soils are eroding and losing their fertility, precious water supplies are being squandered, rangeland overgrazed, forests destroyed, and fisheries overexploited. The heavy use of pesticides has caused severe problems. There is growing human morbidity and mortality while, at the same time, pest populations are becoming resistant and escaping from natural control. In the intensively farmed lands of both the developed and developing countries, heavy fertilizer applications are producing nitrate levels in drinking water that approach or exceed permitted levels, increasing the likelihood of government restrictions on fertilizer use.

Other agricultural pollutants have the potential for damage on a much larger scale. While industry is often to blame, agriculture is becoming a major contributor to regional and global pollution, producing significant levels of methane, carbon dioxide, and nitrous oxide. Natural processes generate these gases, but the intensification of agriculture in both the developed and developing countries has increased the rates of emission. Individually or in combination, these gases are contributing to: acid deposition, the depletion of stratospheric ozone, the buildup of ozone in the lower atmosphere, and global warming. The effects on the natural environment and on human well-being are well known, but in each case there are significant adverse effects on agriculture. In relation to global pollution, agriculture is both culprit and victim.

The Doubly Green Revolution

Some argue that lack of food is simply a problem of unequal distribution. If poor people were not poor, they could buy the food they need. This is true, but oversimplistic and not very helpful. There are no signs the world is about to engage in a massive redistribution of wealth.

In theory the industrialized countries could feed the world. However, this would require several hundred million tons of food aid, compared with only about 10 million tons today. Asia alone would require over 240 million tons of aid in excess of their predicted production, if everyone in Asia were to be adequately fed. This would place extraordinarily heavy burdens on both the donors and the recipients. The environmental costs for the developed countries would be high, and for the developing countries the availability of free or subsidized aid in such large quantities would depress local prices and add to existing disincentives for local food production. More importantly, this scenario implies that a large proportion of the population in the developing world would fail to participate in global economic growth.

The practical reality is that the majority of the poor live in rural areas. The only way they can increase their incomes is through agricultural and natural resource development, which means greater rural productivity, generating both more food and more employment.

Implicitly, this recognizes that food security is not a matter solely of producing sufficient food. For the rural poor, food security depends as much on employment and incomes as it does on food production, and agricultural and natural resource development is crucial in both respects.

Food security, so defined, is also a key determinant of family size. The greater the degree of security and the higher the level of their education, the more will women take advantage of new opportunities and plan ahead for themselves and their families. Appropriate agricultural and natural resource development can also significantly contribute to greater environmental protection and conservation. Finally, vigorous agricultural and economic growth can stimulate world trade, providing significant benefits for all countries, developed and developing.

I believe these arguments, when taken together, point to the need for a second Green Revolution, yet a revolution that does not simply reflect the successes of the first. The technologies of the first Green Revolution were developed on experiment stations that were favored with fertile soils, well-controlled water sources, and other factors suitable for high production. There was little perception of the complexity and diversity of farmers' physical environments, let alone the diversity of the economic and social environment. The new Green Revolution must not only benefit the poor more directly but also must be applicable under highly diverse conditions and be environmentally sustainable.

In effect, we require a Doubly Green Revolution, a revolution that is even more productive than the first Green Revolution and even more "Green" in terms of conserving natural resources and the environment. Over the next three decades it must aim to:

- repeat the successes of the Green Revolution
- on a global scale
- in many diverse localities

and be

- equitable
- sustainable
- and environmentally friendly

The complexity of these challenges is daunting, in many respects of a greater order of sophistication than has gone before. Yet, I am an optimist. I firmly believe we can provide food for all in the twenty-first century. But there is no simple or single answer. It is not just a matter of producing more or enough food. If hunger is to be banished the rural poor have either to feed themselves or to earn the income to purchase the extra food they need. This requires a new revolution in agricultural and natural resource production aimed at their needs. And this cannot be achieved by ecology alone or by biotechnology alone, or by a combination of these. It requires participatory approaches as well—involving farmers as analysts, designers, and experimenters. If we can bring all three approaches together, then we can feed the world in a way that is not only equitable but also sustainable.

Gordon Conway
President, The Rockefeller Foundation

This introduction draws on Professor Conway's recent book, The Doubly Green Revolution: Food for All in the 21st Century, *published in 1999 by Cornell University Press, Ithaca, New York.*

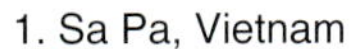

1. Sa Pa, Vietnam
The village of Sa Pa, 1,600 meters above sea level, is about forty kilometers from Lao Cai, a city on the border between Vietnam and China. Access to Sa Pa is along mountain roads. Many members of the H'mong and Zao tribes live in Sa Pa. They plant and harvest rice in terraced fields built like steps in the side of the surrounding hills. At the Sunday morning market, one of the principal enjoyments of the people in the area, this young H'mong boy put his own rice in a bowl of noodles he had ordered, then took his time eating.

2. Calcutta, India
How surprised I was to realize that even today many people in India are very poor. In Calcutta, one of India's best-known large cities, many street people have been chased off to the city's outskirts over the past two years. The city took this action in the hope of attracting foreign investment. In addition, the streets are now cleaned twice a day, in the morning and evening. All such action, however, does not mean that poverty has disappeared. This photo, showing a mother and child early in the morning in busy downtown Calcutta, tells the story vividly.

3. Phnom Penh, Cambodia
Terribly deep scars remain in Cambodia from the long years of fighting. One wonders how the Cambodians, such a gentle race of people, could have been capable of the brutal behavior attributed to them. In fact, the Cambodians themselves probably have the most trouble understanding this. The Russian Market, one of Phnom Penh's principal markets, bustles with people buying food and other items. This visually impaired woman sits with her daughter in the same spot every day, begging for alms.

4. Kathmandu, Nepal
Many important historic buildings line the area around Durbar Square in downtown Kathmandu, the capital of Nepal. In the southwestern section of the square, people come to buy and sell products such as grains, vegetables, spices, and flowers. Here, a young boy and his friend—or maybe they are brother and sister—wait patiently for customers to buy their vegetables. Next to them is a chai (tea) shop, with the shopkeeper preparing for morning customers by boiling tea and milk on a small gas burner. All in all, a typical Kathmandu morning scene.

5. Phnom Penh, Cambodia
Small boats carrying loads of fish caught in the Mekong River gather early in the morning on the riverside in Phnom Penh, not far from a casino boat and one of the capital city's most deluxe hotels. Children come here to pick up small fish dropped from the boats. I wondered whether the children took the fish home to eat with rice or whether they sold the fish to help support their families.

6. Calcutta, India
Calcutta's air is polluted by emissions from motor vehicles with poorly maintained diesel engines, the smoke from factory chimneys, and the smoke from rusty incinerators on wheels used for burning trash in the city. When rain falls during the monsoon season, sewer water backs up onto the streets. Seen here are two barefooted bearers carrying heavy loads on their heads as they trudge along an uneven road.

7

7. Shanghai, China
Shanghai is China's most cosmopolitan city and perhaps the one that foreigners feel most comfortable in. National and local leaders aim to make it one of the largest economic centers in Asia and are aggressively promoting major redevelopment in the city's Pudong district. The area around Yuyuan, a typical residential district for ordinary workers, retains the atmosphere of the old days. On weekends it becomes especially lively. The restaurants are exceptionally good. A poorly dressed mother and child gaze longingly through the window at patrons eating in a well-known dumpling shop.

8. Phnom Penh, Cambodia
Victims of landmines are seen in Phnom Penh eking out a living as beggars. Children sell items such as newspapers and cigarettes. In fact, several hundred children were living on the sides of the roads in the area near the central market, and I wondered what circumstances turned them into street children. I felt uncomfortable taking photos of them.

8

CHICKEN CURRY
Rs.15/-only
मुर्ग कोरमा
15/- रू०

9. Delhi, India
Jami Masjid is a large Islamic mosque located in the center of the old section of Delhi. The area's many restaurants cater to visitors to the mosque. The scene on Chawri Bazaar Street is especially odd. All day long, from early morning to night, one sees groups of men squatting in front of the restaurants quietly waiting. The men sit without moving for as many hours as it takes for someone to give alms to the restaurant owner to pay for food for them, usually nan and vegetable curry.

10. Bhaktapur, Nepal
Bhaktapur, Nepal's ancient capital, is about 10 kilometers east of downtown Kathmandu. Kathmandu, Bhaktapur, and Patan—a city bordering on Kathmandu—have been designated World Heritage Sites by UNESCO. All three have central squares named Durbar in their old city centers. Narrow streets encircle the squares, spreading out net-like. Almost all the homes here are two or three-story brick houses. They characteristically have small wood-framed windows on the second and third floors. One morning passing through this area I came upon a group of men butchering water buffalo meat for sale that day.

Peacock

APTECH
APTECH

11. Dhaka, Bangladesh
Traffic crowds the streets of the old section of Dhaka, the capital, and the central office district. Different from the notorious traffic congestion like in Bangkok, rickshaws comprise an overwhelming percentage of the traffic in Dhaka. These vehicles are an important mode of transportation in the city. Since Bangladesh reportedly has the world's highest population density, that probably makes Dhaka the world's most densely populated city.

12. Mumbai (Bombay), India
Mumbai is India's largest commercial city. Besides being the home of the world's largest movie industry, the city is also the home of just about any other conceivable kind of business activity. Especially interesting for me were the laundry cooperatives and the lunch delivery businesses. Laundry is washed, dried, and delivered the same day. Meanwhile, after hundreds of thousands of wives make lunch at home for their husbands, a delivery service picks up the lunches and delivers them at specific times to the places where the husbands work. The service is reliable and cheap.

13

13. Mumbai (Bombay), India
During the morning rush hour, trains arriving at the Victoria Terminus every five minutes or so from the suburbs of Mumbai are so crowded that passengers hang from their sides. It was also surprising to see commuters—mostly men, but also some women—squatting to relieve themselves, without great concern, on the side of the tracks even when trains were passing.

14. Fatehpur, Rajasthan, India
About 150 kilometers north of Jaipur, the capital of Rajasthan State, rich landowners and merchants have fine homes, called haveli, whose walls are colorfully painted. Haveli owners spend most of their time in large cities. Although many haveli are found in Fatehpur and vicinity, the lives of most of the people living in Fatehpur and other towns around here are far removed from such luxury.

14

15. Lhasa, Xizang, China
Ever since the restrictions on New Year visits to Jokhang Temple, the holiest temple of Lamaism, were lifted during the Tibetan New Year in February 1981, Tibetans began visiting Lhasa not only from places throughout the wide expanse of Tibet but also from faraway Qinghai, Yunnan, Szechwan, Gansu, and Nei Mongol. This photo shows the area in front of the temple's main gate filled with pilgrims wearing heavy sheepskin coats.

16. Baotou, Nei Mongol, China
From about 7:30 a.m., Gangtie Rd. in Baotou is filled with people on bicycles commuting to work at the Baotou Steel Mills. The sixty thousand workers here produce 1.5 million tons of steel a year. As an essential product for China's modernization, steel is produced not only here in Baotou, a town on the Huang He River (Yellow River), but also at large-scale steel mills in Shanghai, Wuhan, Beijing, and Anshan.

17. Manila, Philippines
The Philippines has more Catholics than any other country in Asia. In 1995, Pope John Paul II visited Manila and celebrated outdoor Masses on January 14 and 15. This photo is of the first Mass, with about a million people in attendance. On the following day, a Sunday, over four million people crowded the area around Razal Park, where the Pope said the second Mass.

18. Shanghai, China
On the nights before major holidays, such as National Day (October 1), celebrating the founding of the People's Republic of China, and May Day (May 1), buildings are illuminated in the Waitan (Bund) area along the bank of the Huang Pu River, motor vehicle traffic is stopped, and the main road is thrown open to pedestrians. Several hundred thousand visitors crowd the area to enjoy the brightly lit night scenery. Lights from high-rise buildings and the giant TV tower in the Pudong area on the opposite side of the river add to the scene. Families enjoy eating out at the many restaurants in the area.

19. Chittagong, Bangladesh
Along the coast in Gujarat State, India, and in this area in the outskirts of Chittagong—the second largest city in Bangladesh—there are many scrapyards for breaking huge ships taken out of service. Besides the long beach, ideal for shipbreakers, this area also has a rich supply of cheap labor. Demand is strong for scrap iron and for various fittings taken from the ships. This photo shows workers carrying a steel plate weighing perhaps 300 kilograms removed from the ship in the background. The workers walk barefooted through dirt slippery from oil and grease.

20. Gaya, Bihar, India
Bihar is India's poorest state. Not far from Bodh Gaya, one of Buddhism's holiest places, I came across a spot where thousands of workers were carrying rocks for pulverization by a machine to produce stones for construction use. Women work more than ten hours a day to earn only forty rupees (about one dollar). Even at that pay rate, the women said they felt lucky to have work.

21. Dhaka, Bangladesh
Sadarghat is the central section of old Dhaka. Many workers commute from across the Buriganga River, landing here and then going on to their offices. A scene is thus created each morning and evening where hundreds of small boats vie for space on the river. Because ferry boats coming from longer distances also berth along this same area, often blocking my view, I was continually nervous looking through the camera lens for good shots. The people were obviously used to their commute, though, and many stood calmly in the rocking boats.

22. Darbhanga, Bihar, India
In India, the monsoon season reaches its peak in July and August. When it starts raining during those two months it doesn't stop for hours. Women with no umbrellas carry packages on their heads in the rain and get soaking wet.

23. Paizhouwuan, Hubei, China
Early on the morning of August 1, 1998, during heavy rain, the levee of the Chang Jiang River (Yangtze River) suddenly burst in this area. Forty-four persons lost their lives in the subsequent flooding, including nineteen soldiers of the People's Liberation Army assisting in the area. The buildings of Guangming Elementary School in Zhongbao Village near here were destroyed. Fortunately, all 238 children in the school were unharmed. One teacher, however, the wife of the principal, lost her life. The teachers and the children hope that new school buildings will be constructed quickly.

24. Muzaffarpur, Bihar, India
Floods hit the northern part of Bihar State every year during the monsoon season. Part of the reason for the flooding is that a number of rivers flowing south from Nepal pass through this area to converge with the Ganga (Ganges) River. This area seems to me to have much in common with the northern part of Hunan Province in China. Shown in the photo are friends and relatives of an old man who died of exhaustion during the flooding. The old man's body was wrapped in a blanket and placed on a rough litter on dry ground prior to temporary burial.

25. Samastipur, Bihar, India
The flooding that occurs every year in India's Bihar State and in neighboring Bangladesh is not often reported in the news. In addition, the government provides almost no flood assistance. During flooding, residents of the area carry their household goods (scant few) to the side of the road. The bicycle hanging in the tree here gave me the strong impression that bicycles are among a household's most prized possessions.

26. Paizhouwuan, Hubei, China

The fact that nineteen soldiers of the PLA lost their lives in this area probably means that no one expected the levee to break so suddenly. But extremely heavy rains, reportedly the heaviest in a hundred years, fell continuously, and the flow of water in the Chang Jiang River (Yangtze River) reached an unheard of volume of almost one hundred thousand cubic meters per second. Partly because priority was placed on protecting large urban areas such as Wuhan, Jiujiang, and Yueyang, the agricultural areas along the Chang Jiang suffered tremendous damage. The Paizhouwuan area was no exception.

27. Darbhanga, Bihar, India
Among the persons who suffered from flooding in this area, perhaps those who were able to live temporarily in tents put up in towns and villages were more fortunate than those who could only escape to the sides of roads. Many of the people who fled the flooding took livestock with them. They thus faced the double difficulty of obtaining food for themselves and the animals. Maybe that helps to account for why I saw lines of people here and there leading water buffalo and other livestock toward the marketplace to sell them. They no doubt had the selling prices for their livestock beaten down.

28. Anzhangxiang, Hunan, China
Flooding plagues the northern part of Hunan Province just about every year, caused mainly by the water level rising in the province's four rivers—the Xiang Jiang, Zi Jiang, Yuan Jiang, and Li Shui Rivers. Another cause is land reclamation from the Dongting Hu and Poyang Hu Lakes, reducing the numbers of outlets for water from the Chang Jiang River (Yangtze River). There was also damage to the canal in Anzhangxiang. Here, taking advantage of the agricultural off season, about three thousand farmers were mobilized to repair the damage. Farmers who did not answer the mobilization call were fined thirty yuan (about $3.50) a day.

29. Nanxian, Hunan, China
The flooding damage in Nanxian was also terrible. I talked with Bi Jiajian, whose home was swept away, forcing him and his family to live in a temporary shelter. He's worried about how he's going to get by. He and his wife are old, and he has two sons and five grandchildren to take care of. They all live together, and the only daily assistance the government gives them is food. If Mr. Bi wants to build a new home, he can get five thousand yuan (about $725.00) from the government, but that won't be enough. He himself would have to provide another three thousand yuan (about $450.00).

30

30. Guilin, Guangxi, China
Guilin has a rainy period that begins between the third week in March and the second week in April. About two weeks afterward, the rice farmers in the area begin planting rice in wet paddies. The villages of Yangti (shown here) and Xingping run along the Li Jiang River, and this entire area, together with Gaotian, comprise some of the most beautiful scenery in China. The influence of the beauty of Guilin and the Huangshan Mountains in Anhui Province on Chinese poets and painters is immeasurable.

31

31. Karnal, Haryana, India
Haryana State, once part of Punjab State, is one of India's most fertile districts, especially for producing grains such as rice and wheat. Because over 70 percent of all Indians are vegetarians, rice and wheat are particularly important to them. Rice planting is generally done by hand. Surprisingly, however, I found small tractors being used for cultivating wet rice paddies.

32. Bali, Indonesia
Compared to its relatively small land area, Bali has numerous mountains and valleys. One cannot help but be taken by the combined beauty of the terrace fields, wet paddies, and plush tree growth. The Balinese are the only Indonesians who embrace Hinduism, and their fine arts, classical arts, and religious ceremonies reflect their profound beliefs. As I watched this farmer working a rice field with Mt. Agung, Bali's tallest and holiest mountain, in the background, it seemed for an instant that he was performing a religious ceremony.

33

33. Pagan, Myanmar (Burma)
For about two hundred years from the middle of the eleventh century, the Pagan Dynasty ruled this country and enjoyed great prosperity. During those two centuries, about thirteen thousand Buddhist pagodas and stupas were built along the banks of the Irrawaddy River in the Pagan area. Hundreds of splendid structures from that period are extant to this day. In fact, "Angkor" in Cambodia and the structures in Pagan comprise two of the finest sets of cultural assets in all of Southeast Asia. People living in the Pagan area, however, are poor. They cultivate small plots of arable land and raise mountain sheep. Because of insufficient capital, they are unable to use water from the great Irrawaddy for irrigation.

35

34. Hlegu, Yangon (Rangoon), Myanmar (Burma)
Ever since the military government took control of Myanmar after the coup d'etat in 1962, the country's economy has moved steadily downward. Although formerly the world's largest exporter of rice, for instance, today the United Nations has designated Myanmar a heavily indebted poor country. Although some economic progress is seen in the capital, Yangon, and in parts of Mandalay, the second largest city, the overall feeling is that time stopped in Myanmar several decades years ago.

35. Mandalay, Myanmar (Burma)
Because of irrigation problems, most farmers in Myanmar grow only one crop of rice a year. After enough rain falls during the monsoon season in May, they use cattle and water buffalo to plow the fields and then plant the rice manually, with everyone participating. In the fall, the farmers harvest the rice as delicately as if they were picking up a baby.

36. Bhadrapur, Nepal
The city of Bhadrapur is located at Nepal's southernmost point, just ten kilometers from India's West Bengal State. The climate in this lower part of Nepal, just north of India, is totally different from that in the elevated regions. The soil is relatively fertile, and water paddies are common. Although this photo is of a pastoral scene, not far away are camps holding over one hundred thousand refugees of Nepalese descent from Bhutan. These people are unable to return home and have already been forced to live for ten years in the camps.

38

37. Pyongyang, North Korea (D.P.R.K.)
On April 15, 1992, about fifty thousand students and schoolchildren were mobilized in a mass game to celebrate the eightieth birthday of President Kim Il Sung. The ceremonies were held in a stadium in the capital city of Pyongyang. For the past several years, because of droughts, floods, and other reasons, crops have languished in North Korea, and the food situation is said to be very critical.

38. Lukla, Nepal
Lukla, a village located about 150 kilometers northeast of Kathmandu, Nepal's capital, is encircled by several six-thousand-meter mountains. After a runway was recently completed that let small airplanes fly in and out of here, Lukla became a favorite base for trekkers from around the world who view the Himalayas, especially Mt. Everest. During the peak season in November and December, it is difficult to obtain air tickets. Even during that time, howwever, the sherpas continue to take their yak out and work their farmland.

39

39. Lai Chau, Vietnam
Lai Chau, a town in the mountainous northernmost part of Vietnam, was a difficult place to visit until fairly recently because of the war with France, the Vietnam War, and tense border relations with China. Because arable land is scarce for cultivating rice in wet paddies, many farmers here are still involved in traditional slash-and-burn farming on hills. I'll never forget the sight of the young husband and wife working quietly side by side tilling the dry soil.

40. Kathmandu Valley, Nepal
Kathmandu, is located about 1,300 meters above sea level and the immediate vicinity is surrounded by mountains two and three thousand meters high. About 65 kilometers away, meanwhile, are the tall peaks of the Himalayas, some towering as high as seven thousand meters. The peaks are covered with snow all year round. Almost all the mountains in Nepal up to three thousand meters in height have terraced fields carved into them for raising various crops. Viewing the fields from a helicopter impressed me anew with the tremendous efforts the local farmers make to produce food.

41. Jiaxian, Shaanxi, China
Jiaxian is located about three hundred kilometers northeast of Yan'an. In this area is a large plateau of loess, yellowish-brown loam mainly carried from the northwest region of China by the wind and deposited here. The Huang He River (Yellow River) flows nearby, but a lack of facilities prevents use of the river's water for irrigation. Crops cultivated here on the terraced fields are mainly potatoes, corn, and a little wheat. Because the land is so arid, a successful harvest depends greatly on rain. The farmers are poor but very proud of the historic role Yan'an played in the revolution.

42. Hukou, Shaanxi, China
"He who controls the waters of the Huang He River (Yellow River) controls China." This saying has been borne out throughout China's long history. The saying also emphasizes how the Chinese people have continuously suffered from the river's wildly erratic flow. Upstream, like in Qinghai Province, a number of dams have been built, resulting in much less flooding than previously. Here in Hukou, situated between Shaanxi and Shanxi provinces, is the only falls along the entire Huang He River. Until fairly recently this area had been off limits to foreign visitors.

43

43. Sandouping, Hubei, China
Situated fifty kilometers east of the Three Gorges Dam construction site, the city of Yichang is the site of the head office of the Chang Jiang (Yangtze) River Three Gorges Project Development Corporation. Between Yichang and Sandouping, where the dam is being built, a road was made exclusively for transporting machinery, materials, and construction workers. Several thousand workers put in three shifts a day. Providing food for so many workers is a tremendous task.

44. Sandouping, Hubei, China
Construction of the world's largest dam, the Three Gorges Dam, began in 1993. It is scheduled for completion in 2009. In building the dam, one hundred million cubic meters of rock will be drilled and moved. Plans call for generating 18.2 million kilowatts of electricity per hour, with an artificial lake storing over 39 billion tons of water. This will drastically alter the landscape in the Three Gorges area. Drilling of a hill of tough granite continues, and a lock is being built to allow 10,000 ton ships to pass as far upriver as Chongqing.

44

45. Turpan, Xinjiang, China
Turpan was once an important stop along the Silk Road. When the Buddhist monk Sanzang passed along the foot of Mt. Huoyan Shan (Flaming Mountain) here, he is said to have remarked, "In this sea of fire not a blade of grass grows for eight hundred ri (about 400km) in all directions." The maximum temperature in the summer is 50 degrees Centigrade, and the temperature in the winter drops to minus 30 degrees. The basin here is 154 meters below sea level, making it the second lowest spot on earth after the Dead Sea. It almost never rains in Turpan.

46. Turpan, Xinjiang, China
The Gobi Desert stretches across Nei Mongol, Gansu, and Xinjiang. Here in Turpan, it is called the "Great Gobi Desert," to differentiate it from other parts of the desert. For hundreds of kilometers, fist-size rocks cover the ground like a carpet. It is impossible to cultivate the soil, but some say valuable resources are buried underground in many places.

47. Turpan, Xinjiang, China
Even in the completely dry Gobi some people are seen digging wells. A flow of underground water from the snow melting in the Tianshan Mountains is sometimes found at depths of about forty meters. Streams of water called karez, resulting from several flows of underground water merging together, are numerous in the Turpan basin. They are a vital source of water for the people living in the area.

48

48. Hotan, Xinjiang, China
Hotan is a town squeezed between the Taklimakan (Takla Makan) Desert to the north and the Kunlun Mountains to the south. The Moslem Uygurs comprise 97 percent of Hotan's population. The Uygurs are mostly farmers, but they raise sheep and other livestock as well. Some water flowing from the mountains is found for about 100 kilometers north of Hotan. The Uygurs use this water to eke out a living here in this desert area. Many poplar trees grow in Xinjiang, offering natural protection against desert storms. The Uygurs use their wood as construction materials and for fuel.

红灯 仃车

49

50

49. Anshan, Liaoning, China
Even before the founding of the PRC, Anshan was a key area for the steel industry. The steel plant located here is the largest in China, comprising ten blast furnaces and thirteen rolling mills. But it lacks the modernized facilities the plants in Paoshan (in the outskirts of Shanghai) have, and thus has fallen behind in productivity and the introduction of anti-pollution measures. The annual consumption of coal in China, meanwhile, is approximately 1.45 billion tons, with much of it having high sulfur content. The ill effects of pollution caused by using this coal have become a serious problem not only for the hundred thousand workers at the Anshan Steel and Iron Works but for the entire country and beyond.

50. Daqing, Heilongjiang, China
In the winter in the cold northeastern part of China (former Manchuria), fishermen fish on rivers and lakes by opening holes through ice one to two meters thick. They feed nets through the holes to catch prized winter fish. These fish were once also numerous in the Songhua Jiang River, but because of water pollution they have almost disappeared there. Here in the manmade lake in Daqing, the catch includes carp, grass carp, and catfish. At temperatures reaching minus 40 degrees Centigrade, the fish freeze almost the instant they are pulled from the water.

52

51. Xilinhot, Nei Mongol, China
Winter in Nei Mongol is bitter cold. Most of the people living in the general area of Xilinhot, the central part of the Xilin Gol Steppe, are Mongolian herdsmen. During the winter months, temperatures reach minus 30 degrees Centigrade. The herdsmen set their livestock free to feed on dead grass they find under the snow. But if snow accumulates to heights of fifty centimeters or more, the animals cannot dig the grass out and may starve to death.

52. Xilinhot, Nei Mongol, China
The grass that grows in Xilinhot, "pasture green grass," is considered to be among the highest quality grazing grass in China. The meat from sheep that fed on this grass is also considered to be of prime quality. In fact, the only mutton provided to emperors during the Qing Dynasty was produced here in Xilinhot. Nei Mongol can provide the attractive scene of sheep being milked in the background of a beautiful green prairie.

53

53. Panipat, Haryana, India
A chance encounter put me together with a group of herdsmen from Rajasthan State. It was the dry season, and the herdsmen were on the move looking for good grazing for their cattle. The group comprised about ten families. They were herding several hundred cattle, living from place to place as they moved. Included among the cattle were many calves. Perhaps the herdsmen earned a living as they moved by selling milk from the cows. I noticed some children in the group, and knew there was certainly no way for them to be attending school.

54. Wamena, Irian Jaya, Indonesia
The flight from Jakarta to Jayapura, the capital of Irian Jaya Province, takes nine hours. Another hour southward aboard a small propeller-driven cargo plane can then carry you to Wamena, a small town encircled by four-thousand-meter mountains. Peoples of the Dani, Lani, and Jali tribes inhabit this area. They cultivate crops such as sweet potatoes, corn, and peanuts. Traditionally the men wear very little clothing. Many of those who work the fields wear only a sheath called a kotega.

55. Sa Pa, Vietnam
Members of the H'mong and Zao tribes visit the Sunday market in Sa Pa attired in native dress. Both men and women of the H'mong wear costumes dyed a dark blue. The clothes are made at home and colored using natural dyes. In this photo, the children's hands have blue splotches from helping with the dyeing. Sugar cane is a special children's treat on Sundays.

56. Naisoi, Mae Hong Son, Thailand
The members of the Padaung Tribe in Kaya State in Myanmar (Burma) traditionally engage in fa-ming. For more than ten years, however, several hundred members of the tribe have been living as refugees after having escaped to Thailand from Myanmar because of the military government there. The Padaungs are not allowed to farm in Thailand and live on income earned from tourists attracted to the women of the tribe, called "giraffe women." From childhood, these women wear layers of brass rings around their necks. Indeed, Padaungs are also called the "long-neck" tribe.

57. Bali, Indonesia
Balinese bury their dead soon after they die. Every ten years or so, however, each village exhumes the bodies and cremates them in a joint ceremony. The participants, all wearing mourning clothes, form circles around the remains of their family members, and then eat the food they brought with them and drink to the spirit of the deceased. After cremation, the families carry the ashes to a nearby river or to the sea; while offering Hindu prayers, they scatter the ashes on the water.

58. Wamena, Irian Jaya, Indonesia
The people of Wamena eat a special dish called masak at weddings and funerals. Masak is prepared by baking pork and sweet potatoes together. This photo shows family members and relatives sitting around eating masak after the funeral service for an old man in the family. Except for some food grown locally, almost all items needed for everyday life are flown into Wamena on cargo planes. Until a new road is completed several years from now, Wamena will remain an inland island isolated from other communities. The ordinary food of the people of Wamena is extremely simple.

59. Taunggyi, Shan, Myanmar (Burma)
One of the attractions of Taunggyi in Shan State in Myanmar is the market held every five days. Because the town is situated on a plateau 1,800 meters above sea level, even in summer it is cool enough in the early morning and evening to require wearing a long-sleeve shirt. The market attracts the members of tribes living in the nearby mountains and takes on a unique atmosphere. The Shan are of Chinese lineage. For that reason, and because they live in an area bordering on China, they have always had close relations with people on the Chinese side. This area is also infamous worldwide as a production center for illegal drugs.

60

60. Inle, Shan, Myanmar
The members of the Intha Tribe in Shan State live on Inle Lake. Because they live on the water and thus don't get much exercise, the men of the tribe are said to have highly developed foot muscles for manipulating their boat paddles. The Intha even grow vegetables in fields floating on the lake. The Phaung Daw U Pagoda on the lake is also famous. All boys in Myanmar must undergo a period of Buddhist training, even if only for a short time. Although young, the boys are forbidden to eat anything between noon of one day and the next morning.

61. Amritsar, Punjab, India
Amritsar is famous as the site of the head temple of Sikhism. Sikhs are deeply religious and many of them succeed in the military and in business. The Main Hall of the head temple is covered with gold leaf, making it literally a "Golden Temple." Marble covers the grounds of the temple and not a speck of dirt is found anywhere. All visitors to the temple are given free meals; preparing thousands of meals a day is one of the services performed by the Sikh adherents.

61

62

62. Tawitawi, Philippines
Tawitawi is the name of the southernmost islands in the Philippines. Many inhabitants and those of Mindanao and the Sulu Islands to the south, are Muslims. The standard of living is quite low. The town of Bongao, with its nearby airport, is the center of Tawitawi. Almost all the children do some kind of work.

63. Ben Tre, Vietnam
The most affluent area in the Mekong Delta is Can Tho Province. Farmers in Can Tho harvest four rice crops a year. Ben Tre, however, although located fairly near Can Tho, is a poor area. Salt water flows upstream from the ocean along Ben Tre so that rice can only be harvested once a year, during the monsoon season, when the water level of the Mekong River is high. On the other hand, coconut trees are hardy in this climate, and production of cakes called mutdua, which use coconut as an ingredient, is brisk.

63

64. Kashgar, Xinjiang, China
Kashgar is China's westernmost city, located about five thousand kilometers from Beijing. It is also the cultural center of the Uygurs. There is a romantic air to Kashgar, once an essential stop along the Silk Road that enthralled people like Marco Polo and the Swedish explorer Sven Hedin. There are many peddlers along the sides of the road in the old section of Kashgar. The woman here is selling boiled eggs, colored red.

65. Kashgar, Xinjiang, China
The back streets of the old section of Kashgar are mazelike, though the type of stone used in a road tell you, for instance, whether it is a dead end. A feature of the homes in the old city is that each has a small inner yard. A carpet covers the living room floor inside the houses, but there are no seats or sofas. People sit on the floor, stretching their legs, and either eat or join in conversations. A peddler selling fruits and vegetables passes quietly along one of the typical alleys in the old section of Kashgar.

66. Kashgar, Xinjiang, China
In the center of the old section of Kashgar stands the Id Kah Mosque, built many centuries ago. For the first prayers on Friday afternoon several thousand Muslims assemble. No women are allowed inside the mosque. Around the time the prayers end, women wrapped in chador carry nan and water to the mosque entrances and wait outside for the men to emerge. The men, freshly purified from their prayers, are asked to breathe gently on the nan and water. Doing so is said to bring good health, happiness, and prosperity to those who receive this blessing.

67

67. Lhasa, Xizang, China
Butter made from the milk of yaks is an essential food for Tibetans. Besides butter—called ma in Tibetan—cheese and yogurt are also made from yak milk. Drinking tea with lots of ma in it and eating zanba, a type of meal made from rye flour, is the favorite food of Tibetans. Many shops in the Octagonal District in the old section of Lhasa are kept busy selling ma.

68. Bhaktapur, Nepal
Mornings during the winter throughout the Kathmandu basin are often foggy, turning Bhaktapur, with its rows of buildings seemingly from the Middle Ages, as well as Kathmandu and Patan, into fantasy cities. After the fog dissipates, the sun emerges to bring out the bright colors of the fruits and vegetables that vendors are selling, but they have few customers.

69

69. Jaisalmer, Rajasthan, India
The area around Jaisalmer, on the western edge of India, near the border with Pakistan, has many military installations. The city, however, seems to date from the Middle Ages. Vegetable markets in the city do a brisk business, and sacred cows are apt to appear suddenly from any direction.

MING LI ONG
KAYO BAY MAY SULIRANIN?
ALAMIN ANG KAPALARAN?
Mde. BETH T. ONG
THE CONTROVERSIAL STORY OF THE 200,000 JAPANESE SEX SLAVES OF WORLD WAR II
JOEL TORRE
RICARDO CEPEDA
SHIRLEY TESORO
and
SHARMAINE ARNAIZ
COMFORT WOMEN
A Cry for Justice
ALYSSA FILMS
Released by FOUR N FILMS
Directed by CELSO AD. CASTILLO
VIVA FILMS presents
BING LOYZAGA
ANJANETTE ABAYARI
GARY ESTRADA
&
RUSTOM PADILLA
BRAT PACK
PAMBAYAD ATRASO
Directed by: DEO J. FAJARDO, JR.
WHOLESALER · RETAILER

71

70. Manila, Philippines
Chinatown is one of the business centers of Manila, and Tondo is the city's largest slum area. Recto Avenue, which runs between the two areas, is one of the liveliest shopping streets in Manila. Daily necessities are sold at cheap prices. Gaily decorated jeepneys and horse-drawn carriages add to the general congestion in Chinatown from early morning into late evening.

71. Calcutta, India
A trash dump on a corner lot in the city draws crows and ragpickers. Scenes like this are not uncommon in many Asian cities. On the other hand, some people in the same cities live in great luxury.

72

72. Guangzhou, Guangdong, China
People in Guangdong are famous for their gourmandism. Dog meat, in particular, eaten not only in Guangdong but widely throughout China, is a winter delicacy. Koreans are also said to be great dog meat fans. My Chinese friends say it warms the body in winter and is quite delicious.

73. Guangzhou, Guangdong, China
All sorts of food products are sold at the market on Qingping Road in Guangdong's old downtown area. Many Cantonese are proud of the saying applied to them, that the only things they won't eat that fly are airplanes and the only four-legged things they won't eat are tables. The store in this photo is not a pet shop. Sold here for food are not only dogs but also snakes, cats, monkeys, and even owls. Snake and cat meat eaten together is called "dragon and lion" food.

74

74. Hotan, Xinjiang, China
Every Sunday a large bazaar is held in Hotan. Items on sales include livestock (sheep, cattle, camels, etc.) and food products such as meat (some pork), as well as daily necessities. Especially popular with visitors are noodles, eaten at outdoor food stalls. This shop had the heads and feet of cattle lined up outside. I wanted to ask how the heads are prepared for eating, but before I could find out I unfortunately had to leave the bazaar area to take other photos elsewhere.

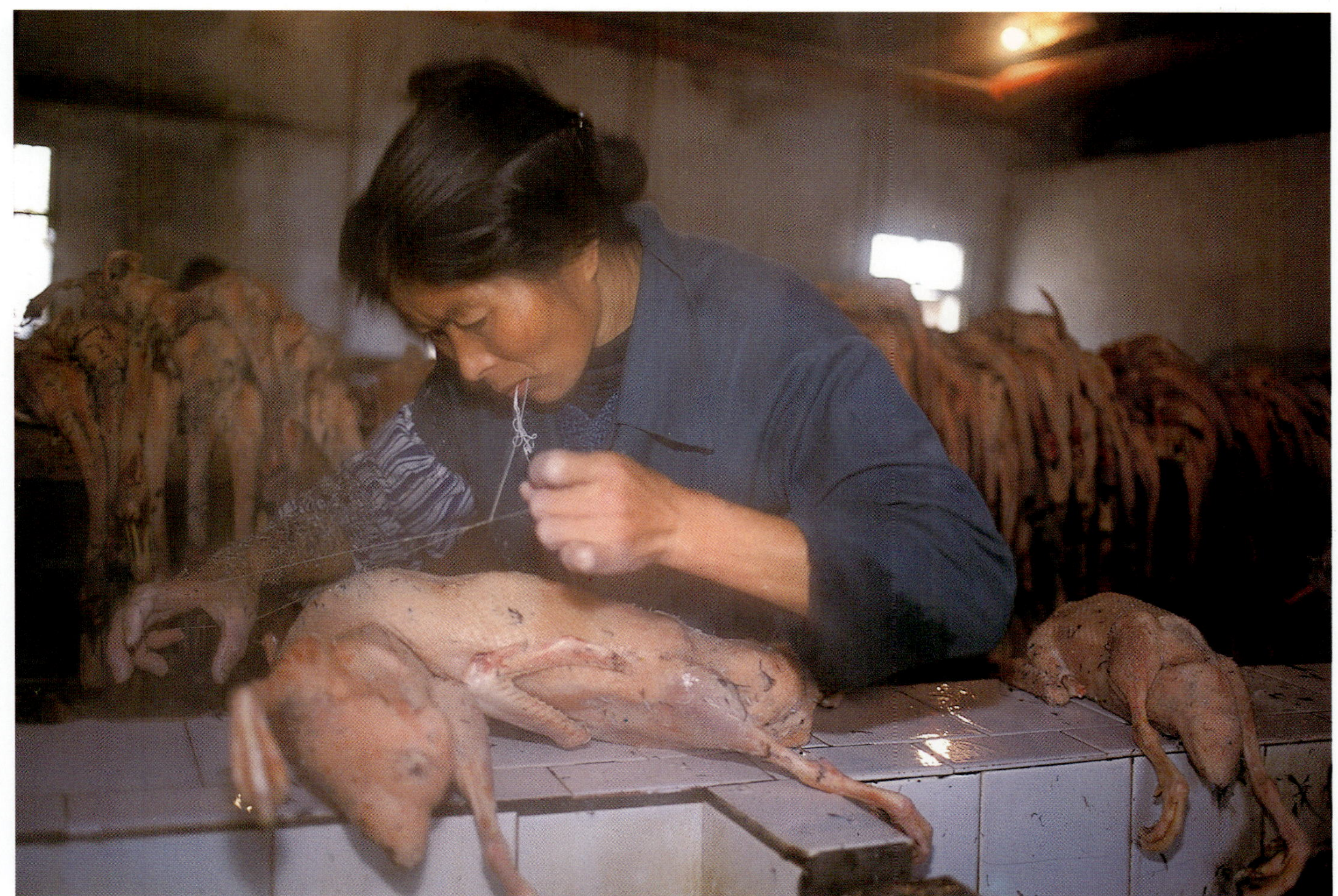

75

75. De'an, Jiangxi, China

Together with Sandong Province, Jiangxi Province is China's center of duck farms. The ducks used for preparing the famous Beijing duck dishes have white feathers, while almost all the ducks prepared in Jiangxi Province have brown feathers. Here at the Gong Qing Ban Duck Farm, seventy workers process an average of 1,500 ducks each day. About forty of the workers are women. Their job is to pluck the final pieces of feather from the ducks' skin. The women ingeniously hold a piece of twine in their mouth, keeping both hands free to wind the twine around the remaining pieces of feather quills and pluck them. Although their base pay is low, the women also receive 0.5 yuan (about $.07) for each duck they process. A skilled woman can process eighty ducks a day.

76. Hotan, Xinjiang, China
From Hotan, I rode in a four-wheel-drive vehicle and headed for the Kunlun Mountains. About ten minutes after leaving Hotan proper we were already in the mountains, and there wasn't a single tree to be seen. That surprised me, for I had imagined the mountains to be covered with trees. We drove along a deep valley, and the road headed inexorably toward higher mountains. The wind grew stronger, and I had an eerie feeling. We passed a jail and the road to a mine along the way, and eventually reached a pass. Directly below us on the pass's opposite side was a village built around an oasis. Called Kashi Tashi, about 4,800 Uygurs live here. For lunch we ate freshly baked nan.

77

77. Jaipur, Rajasthan, India
Castle walls enclose the old city section of Jaipur, the capital of Rajasthan State. Because the castle walls and the walls of the homes in Jaipur are painted pink, the city is also called the "Pink City." Near Ghat Gate a market opens from early morning where milk (mainly from water buffalo) is sold. Persons influenced by Indian culture like to drink strong tea, usually a number of cups a day, with lots of milk and sugar in it.

78

78. Calcutta, India
Early morning in Calcutta sees the food stalls on the streets preparing for the day's business. Many young boys work in these stalls. Besides preparing food, they also make deliveries and do the cleaning up, working right through until late at night. They sleep on stools they line up by the stalls or in a corner at the side of the street. They get almost no time off, and I imagine they get paid very little.

79. Phnom Penh, Cambodia
Brisk activity marks mornings near the central market. The citizens of Phnom Penh enjoy a breakfast of noodles and Chinese tea or delicious French bread and coffee. That is how busy days begin for young sleepy-faced shop helpers. Pretty soon the children of the shopkeeper will go off to school in a three-wheeled cyclo, the shop helpers gazing longingly after them.

79

Canon

80. Kashgar, Xinjiang, China
Because Kashgar is a dry area, water is a precious commodity. The volume of water that flows from the melting snow in the Kunlun Mountains is gradually decreasing every year in Kashgar and Hotan, causing unease among the populace. In the old city of Kashgar, people must go to places like that seen in this photo to obtain water for daily living needs. Walking around selling water in the city would be a good business.

81. Phnom Penh, Cambodia
Men and women, old and young alike, work for hours taking the guts out of small fish caught in the Mekong River. Many children don't go to school but help out with this work. Most people's pay is probably just enough to cover their family's basic needs each day.

82. Hon Dat, Vietnam
The development of waterways in the Mekong Delta amazed me. In fact, I wouldn't be surprised now to hear that the total length is no less than ten thousand kilometers. It is possible to travel anywhere in the delta, even from Ho Chi Min (formerly Saigon). Because of the waterways, duck farms have sprouted up in many places. The delta area supplies almost half of Vietnam's total demand for food.

84

83. Can Tho, Vietnam
Can Tho is the largest city in the Mekong Delta. About thirty kilometers south of Can Tho is Phung Hiep, a town with a floating marketplace. The canal is wide in this area, and before dawn a large number of boats converge, carrying people to buy and sell all sorts of products, mainly rice, fish, fruits, and vegetables. The vendors are mostly women. A Vietnamese friend told me that because an increasing number of people were being injured in boat collisions, traffic regulations have been introduced.

84. Dali, Yunnan, China
The town of Dali is situated on a plateau in the northwest part of Yunnan Province. The people here are mainly of Bai nationality. To Dali's west is Mount Diancang, soaring four thousand meters high. To the east is Erhai, a freshwater lake. Fish are bountiful, and one can even see cormorant fishing. Because a strong wind frequently blows down from the surrounding high mountains, Dali is called the "Windy Town." The lake's waters were exceptionally calm on the morning of this photo.

85. Hong Gai, Vietnam
Halong Bay is called the "Guilin of the Sea." It is one of Vietnam's most picturesque spots. Because it is situated just south of the border with China, for a long while it was off limits to foreign visitors. I was told that August is the most beautiful month to visit. A particularly enjoyable eating experience is to boil shellfish in seawater and eat them while cruising among the many islands in the bay aboard a small pleasure boat. Hong Gai is the main port in Halong Bay.

86. Pematangsiantar, Sumatra, Indonesia
With annual production of four million tons, Indonesia is the world's second largest producer of palm oil, next only to Malaysia's annual production of eight million tons. A strong demand exists for palm oil today because it is low in cholesterol. The trees from which palm oil is extracted are between five and twenty-five years old. After twenty-five years the trees are cut down and seedlings are once again planted. Palm tree groves require a great deal of labor. Also, there is growing concern about the random development of palm tree plantations and the negative influence they may have on the region's ecological chain.

A 5

88. Tawitawi, Philippines 88
The ocean in this area has a coral reef, and the water is perfectly clear, making Tawitawi seemingly a paradise. The fishermen, though, frequently fish unlawfully, using dynamite. Also, if they see a good opportunity to do so, they immediately change from being fishermen to being pirates. It is dangerous for foreigners to visit this area unless escorted by Filipino marines, stationed nearby.

89. Bacolor, Luzon, Philippines
Mount Pinatubo had no history of erupting before 1991 when the largest eruption of the twentieth century occurred here. Fertile farmland and a number of old and historic towns in the vicinity were completely buried by lahar, a mixture of volcanic ash and mud. A billion square meters of lahar remain on the mountain, and every time it rains heavily the lahar moves.

87. Manila, Philippines
In November 1996, on the occasion of hosting a meeting of the APEC countries in Subic City, situated over 100 kilometers west of Manila, former President Ramos ordered the forced closing of the huge Smokey Mountain rubbish dump, calling it a "national disgrace." Despite strong opposition from local residents, Smokey Mountain was closed. Although the original Smokey Mountain, a symbol of Asian poverty, thus vanished, other Smokey Mountains exist. This photo was taken in 1995.

87

90. Mindanao, Philippines
Because of the rarity of typhoons hitting Mindanao, the island is an ideal spot for growing bananas. To develop more land for growing bananas, forests have been leveled and large-scale banana farms built. The bananas grown here are chiefly for export to Japan, South Korea, and Hong Kong. To ward off some insects and mold, various agricultural chemicals are sprayed on the bananas. It is said that the local people will not eat bananas destined for export.

91. Tabang, Kalimantan, Indonesia
A tropical rain forest second in size only to that seen in the Amazon basin used to exist in Kalimantan, the southern half of the island of Borneo. The pilot of the plane I chartered laughed wryly and said that if environmental restrictions were not introduced quickly, the rain forest in Kalimantan will disappear in about ten years.

92. Jakarta, Indonesia
Tamrin Road and Sudirman Road running north and south through the center of Jakarta, the capital of Indonesia, are the main streets in the city, and central government offices, banks, office buildings, and five-star hotels line them. Walk down any of the side roads off the two main streets, however, and you'll find open lots where construction projects seem to have been halted. Many poor people use the empty lots to grow vegetables.

Postscript

How many readers of this book know that even in our day and age, as the world enters the twenty-first century, about 35,000 infants lose their lives every day due to malnutrition? When I was first approached over a year ago concerning the possibility of a book of photographs like this, the matter of infants dying from malnutrition was something I had never thought about before, and I was aghast on learning the large number. Because I knew almost nothing about the world's food problems or agriculture, I asked the staff in my New York office to gather basic materials for me to read. That's when I learned about the 35,000 infants. Not long after that, I read the news about the Chang Jiang River floods in China. Without a moment's hesitation, I decided to accept this assignment to look anew at Asia from the perspective of "food."

Realizing there were tight restrictions on the time available for taking photos and interviewing people, I decided to focus mainly on China, the Indian subcontinent, Indonesia, Vietnam, and Cambodia.

In the summer of 1998, the Chang Jiang and Songhua Jiang (in the Northeast) river basins experienced the heaviest rains in a hundred years. The damage from flooding was terrible, affecting two hundred million people. In Jiangxi Province, where a section of Jiujiang was inundated, the flow of water in the Chang Jiang River at its peak was about one hundred thousand cubic meters per second. That computes to 8.6 billion cubic meters of muddy water flowing past the area in one day. The figure is so large, it is difficult to comprehend its significance.

The history of China has also been the history of the Chang Jiang (Yangtze) and Huang He (Yellow) Rivers. There is even a saying that those who control the Chang Jiang and Huang He Rivers control China. Throughout the northern part of Hunan Province, flooding still occurs just about every year. As a step toward preventing flooding from the Chang Jiang, work continues to move forward on building the high-priority Three Gorges Dam in Sandouping, Hubei Province, expected to be the world's largest dam. It would take seventeen years, with completion scheduled for 2009. Because the ecological chain in the area will suffer, and because the landscape and cultural assets around the Three Gorges, which inspired so many Chinese artists and poets over the centuries, will be altered, much criticism has been aimed at this project from both inside and outside China. Nor will it be a simple matter of relocating the roughly one million people who live in areas that will be under water when the dam is completed. Final voting on whether to approve or reject the dam project was held at the parliamentary session of the National People's Congress in the spring of 1992. Voting was 67 percent in favor and about 7 percent against; over 25 percent of the representatives abstained. A person on the construction project proudly explained to me that since the dam will hold up to 39 billion cubic meters of water, it will prevent the kind of flooding experienced during the summer of 1998.

Starvation has been prevalent throughout China's history, even at one time, some people believe, in the years following the creation of the People's Republic. During the flooding of 1998, the government's priority was delivering an emergency supply of food to the afflicted. The speed with which this was accomplished deserves applause. It seemed that the government had grasped the deepest concerns of the people.

In sharp contrast to the areas afflicted by flooding, I also visited Xinjiang, the loess plateau in Jiaxian, and other arid parts of China. At such places I was keenly reminded of how severe and unmerciful nature can be. China has tremendously wide expanses of land so high or so arid that is it just about impossible to cultivate it as farmland. They include the Tibetan Plateau, over four thousand meters high, and the Taklimakan and Gobi deserts. A comparison of per capita arable land among selected countries shows China with only 1/30th the arable land of Australia, 1/20th that of Canada, and 1/9th that of the United States. The situation is much the same in other Asian countries—and 60 percent of the world's population lives in Asia.

Although the great number of people in China is overwhelming, a trip to India is apt to drive one to despair. Besides a great disparity between the rich and the poor, there is such a very large number of poor—no, not just poor, but unbelievably poor—people in India. My visit there this time was the first in ten years. Besides large cities such as Calcutta, Mumbai (Bombay), and Delhi, I also visited the countryside in several states, including Rajasthan and Bihar. From what I saw, I can only say that compared to my previous visits the number of poor people appears to have increased radically.

Reports say that India's population has just reached one billion and is increasing at the rate of another 45,000 persons every day. That rate means an annual increase equivalent to the size of Australia's total population. It is a fact as well that in order to contribute toward the support of their families a large number of children in India do not attend school and are forced to work long hours. Their parents are poor and have to depend on what the children can earn. In that situation, the more persons working in a family the better, one reason why couples have many children. This in turn ties to an increase in illiteracy, a vicious cycle.

The solution to the issue of the population increase is immensely difficult. My feeling, though, is that in the final say the only solution is education. To ensure that children receive an education, for example, I wonder if it isn't possible, depending on need, to develop a system of paying children to attend school up through the compulsory education level. Some may say there are not enough teachers, schools, or funds. A pressing issue is that many children have to help with work at home.

Wherever I traveled in Nepal, China (Hotan and the loess plateau in Jiaxian) Vietnam, Indonesia, and other countries, whether in mountains, hilly land, or desert regions, I felt that the people have already cultivated all the land capable of cultivation. But I also saw many farmers who even today can't get enough water to meet their daily needs and who have no electricity. All they can do is continue working, buy their daily necessities with the little income they earn, and somehow live their lives. Life for minority groups is particularly severe, and in some countries the landlords take much of the crops the farmers harvest.

The economic growth in East Asia in recent years has attracted worldwide attention and has been the target of envy of less prosperous countries. In becoming preoccupied with the idea of "progress," however, we have overlooked important issues such as agriculture, poverty, population, the destruction of nature and environmental pollution. Much of the available food is becoming unsafe due to the abusive use of chemical fertilizers and agricultural pesticides. In addition, many people are tormented by government policy, outworn customs, discrimination, violence, and corruption. The burgeoning of urban centers, overdensity, and a worsened environment due to development all contribute to the food shortage.

The more affluent a country is in Asia, the less it seems able to provide food to satisfy the needs of its own people. Japan, for instance, supplies just 28 percent (1997) of its needs for grain, the lowest such percentage among the world's seven most advanced nations. (For comparison, the same figures are just under 140 percent for the United States, over 180 percent for Canada, 200 percent for France, 130 percent for the U.K., and just under 120 percent for Germany.) South Korea, meanwhile, is a little over 30 percent. In Singapore, because everyone lives in an urban environment, the figure is probably less than 1 percent.

The food situation in my own country, Japan, concerns me greatly. Fortunately, Japan produces more rice than its populace consumes, and even has reserves of rice. However, Japan produces only 3 percent (1997) of the soybeans it needs, one of our most essential foods.

Countries should be able to produce at least 50 percent of the food they need. It is also wise to build a six-month reserve of food, as some European countries have done. Construction of facilities for storing food reserves, however, requires huge layouts of capital. Many of the developing economies don't have that much extra capital, and thus large quantities of precious food are being wasted. In part, because of imports, the Japanese people today have more than enough to eat, although a great amount of food is being wasted as loss incurred in the distribution process and as leftovers from meals. On the other hand, close to one billion people throughout the world are suffering from a chronic lack of nutrition. Can't something be done about this situation?

Food products have today become a part of the market economy mechanism. If money is available, therefore, it is possible to import any kind of food at anytime from anywhere. Some developing economies export food products as a means of obtaining hard currency. If a food product can be sold more dearly by exporting it than by selling it domestically, then naturally, based on market principles, the emphasis will be on exporting. However, while exports contribute to the country's economy, the supply of food for domestic consumption decreases, and domestic prices thus increase. Those who suffer the most, of course, are the poor.

Even though countries with wealth can import the food they need, their security is an illusion. War or poor weather may cut off the food supply from producing countries. A nonprofessional like me can't help thinking there is much waste in the fact that food products are handled completely as general products and made to depend on the market economy. It would seem acceptable for at least 50 percent of the volume of food consumed in a country to be handled apart from the market economy. I wonder, too, if it isn't possible to establish something like a "world food bank." The World Food Program (WFP) and the Food and Agriculture Organization (FAO), United Nations agencies that do not have food reserves of their own, must spend much effort searching for countries to help with food when they move to assist countries faced with food problems. And almost without exception the matter always turns political, and it becomes difficult to provide the assistance expeditiously. Although the operation of a world food bank would also be extremely difficult, I dream of a capital reserve of about one hundred billion dollars and a food reserve of about one hundred million tons. The organization would establish offices throughout the world and have tens of thousands of food reserve facilities in the world's food production centers. Then, when a need arises, the organization would be ready and able to distribute food promptly. Also, if 1 or 2 percent of the military budgets of countries around the world could be set aside for resolving the world's food problems, many millions of people would be saved.

In taking the photos and conducting the interviews for this book, I came to see up close what it means for large numbers of people to labor long hours doing arduous work to earn even a small portion of the daily food needed for themselves and their families. It truly pained me to have to turn my camera toward those people, and toward people who don't even have that kind of work and must beg for a living. But we must not turn away from this reality. The twenty-first century must not be a century in which people die from malnutrition. In order to produce safe food in sufficient volume it is important to take care of the world's farmland and the persons cultivating it. Without a solid agricultural base, mankind is not apt to have a tomorrow.

September 1999
Hiroji Kubota

Acknowledgments

The photographer wishes to extend his deep gratitude to the following :

Kofi and Nan Annan
Asia Society
Lester Brown
Janet Byrne
CANON
Anna Cataldi
Eleen Cheung
China Cultural Int'l Tours
Contrasto
Gordon Conway
DNP
Doi Technical Photo
Tom Elliott
FAO
Comune di Firenze
FUJIFILM
Kensaku Hogen
IAO
ICP
Ienohikari Kyokai
JA Group
Jim Mairs and Gina Webster
Dom Moraes
NHK Joho Network
NIKKEI
PGI
Art Presson
Katsuhisa and Mamiko Uchida
UNEP
UNFPA
UNHCR
Kerry and Sue Wark
White Space
Xinhua News Agency